~A BINGO BOOK~

General Science Bingo Book

COMPLETE BINGO GAME IN A BOOK

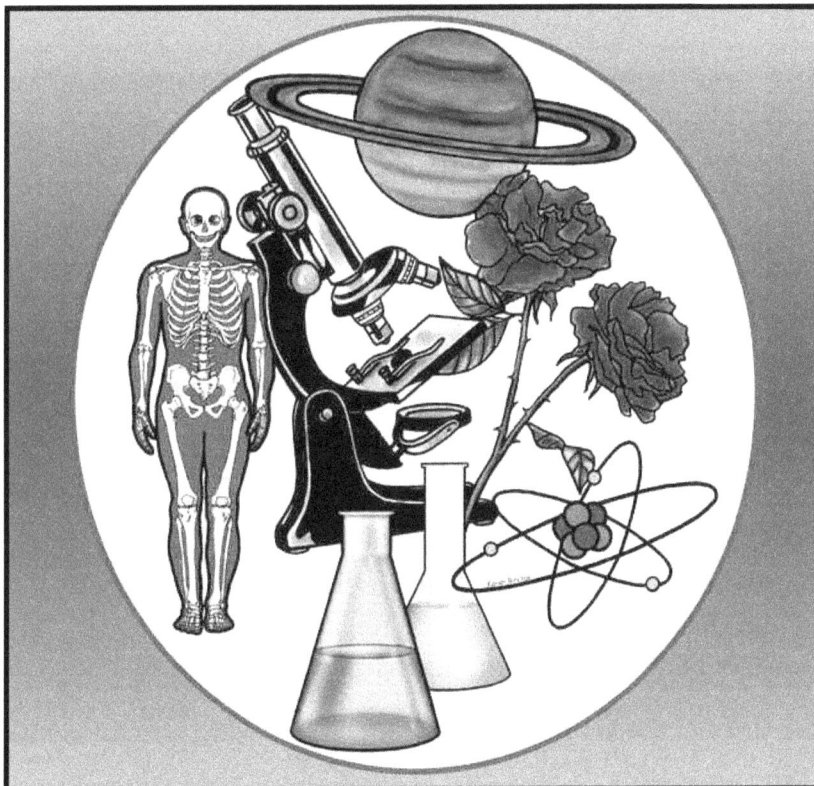

Written By Rebecca Stark

Educational Books 'n' Bingo

Educational Books 'n' Bingo

ISBN 978-0-87386-444-2

Printed in the U.S.A.

GENERAL SCIENCE BINGO DIRECTIONS

INCLUDED:

List of Terms

Templates for Additional Terms and Clues

2 Clues per Term

30 Unique Bingo Cards

Markers

1. **Either cut apart the book or make copies of ALL the sheets. You might want to make an extra copy of the clue sheets to use for introduction and review. Keep the sheets in an envelope for easy reuse.**

2. Cut apart the call cards with terms and clues.

3. Pass out one bingo card per student. There are enough for a class of 30.

4. Pass out markers. You may cut apart the markers included in this book or use any other small items of your choice.

5. Decide whether or not you will require the entire card to be filled. Requiring the entire card to be filled provides a better review. However, if you have a short time to fill, you may prefer to have them do the just the border or some other format. Tell the class before you begin what is required.

6. There are 50 terms. Read the list before you begin. If there are any terms that have not been covered in class, you may want to read to the students the term and clues before you begin.

7. There is a blank space in the middle of each card. You can instruct the students to use it as a free space or you can write in answers to cover terms not included. Of course, in this case you would create your own clues. (Templates provided.)

8. Shuffle the cards and place them in a pile. Two or three clues are provided for each term. If you plan to play the game with the same group more than once, you might want to choose a different clue for each game. If not, you may choose to use more than one clue.

9. Be sure to keep the cards you have used for the present game in a separate pile. When a student calls, "Bingo," he or she will have to verify that the correct answers are on his or her card AND that the markers were placed in response to the proper questions. Pull out the cards that are on the student's card keeping them in the order they were used in the game. Read each clue as it was given and ask the student to identify the correct answer from his or her card.

10. If the student has the correct answers on the card AND has shown that they were marked in response to the *correct questions,* then that student is the winner and the game is over. If the student does not have the correct answers on the card OR he or she marked the answers in response to *the wrong questions,* then the game continues until there is a proper winner.

11. If you want to play again, reshuffle the cards and begin again.

Have fun!

TERMS

ACIDS	HEAT
ANATOMY	HERBIVOROUS
ASTRONOMY	LIGHT
ATMOSPHERE	MAGNET
ATOMS	MAMMALS
BIOLOGY	MANTLE
CARNIVOROUS	MASS
CELL	MATTER
CHEMISTRY	MICROSCOPE
COMPOUNDS	MOLECULE
CORE	MOTION
CRUST	NUCLEUS
DATA	OCEANOGRAPHY
EARTHQUAKES	PHOTOSYNTHESIS
ECOLOGY	PHYSICS
ELECTRICITY	PLANET
ELEMENT	RADIATION
ENERGY	SCIENTIFIC METHOD
EROSION	SOLAR SYSTEM
EXPERIMENT	SOUND
FOOD CHAIN	TEMPERATURE
FREQUENCY	THEORY
GENE	TIDES
GEOLOGY	VOLCANOES
GRAVITY	WATER CYCLE

Additional Terms

Choose as many additional term as you would like and write them in the squares. Repeat each as desired. Cut out the squares and randomly distribute them to the class. Instruct the students to place their square on the center space of their card.

General Science Bingo

Clues for Additional Terms

Write three clues for each of your additional terms.

_____

1.

2.

3.

_____

1.

2.

3.

_____

1.

2.

3.

_____

1.

2.

3.

_____

1.

2.

3.

_____

1.

2.

3.

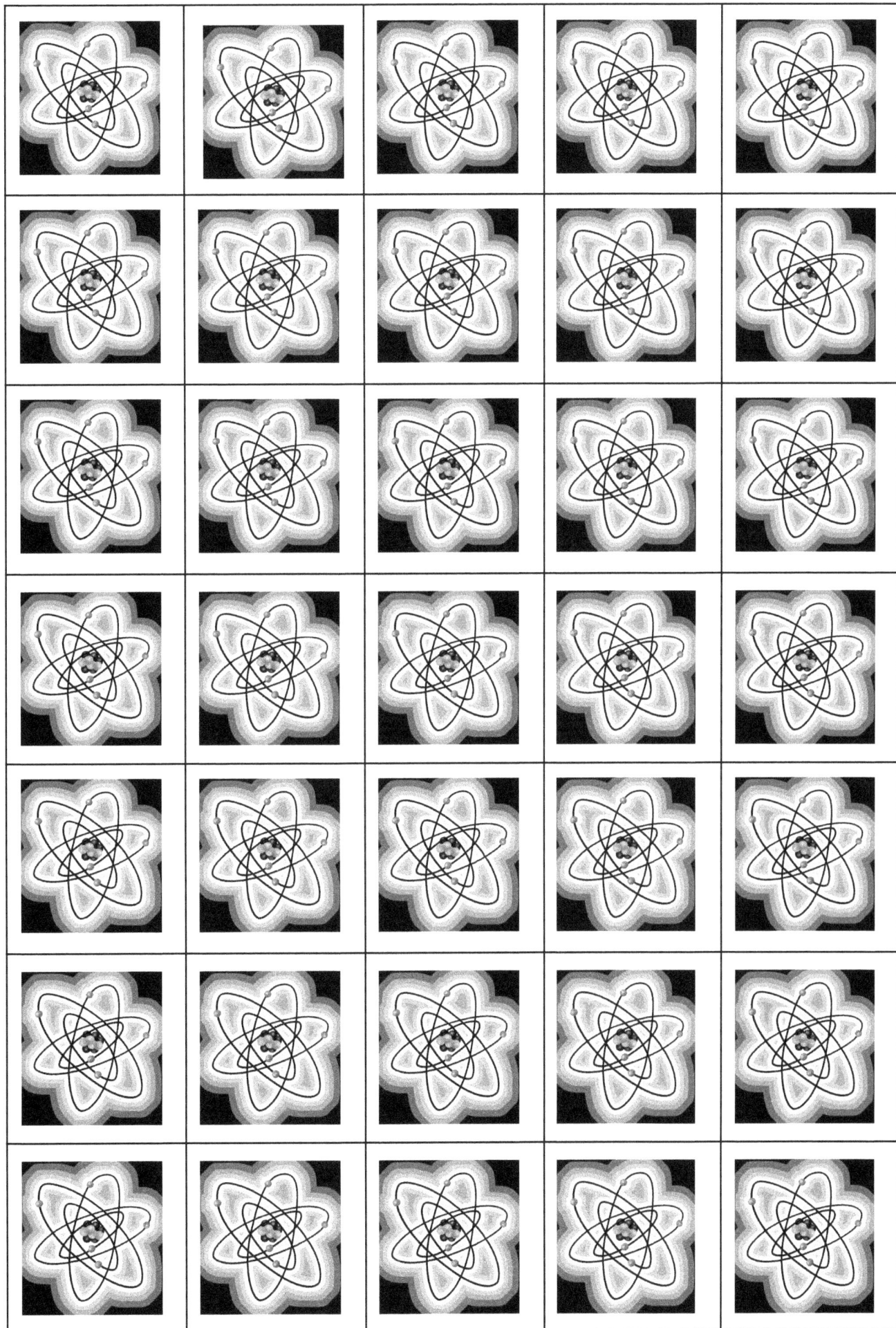

ACIDS
1. They have a sour taste.
2. They turn litmus paper red.
3. These substances yield hydrogen ions when dissolved in water.

ANATOMY
1. It is the study of animal form, especially the human body.
2. It is the science of the shape and structure of organisms.
3. This subdivision of morphology deals with the structure of animals. (Morphology is the branch of biology that deals with the form and structure of organisms without considering function.)

ASTRONOMY
1. This is the study of celestial objects.
2. It is the study of objects and matter outside Earth's atmosphere.
3. This branch of science would include the study of the planets.

ATMOSPHERE
1. This is the name we give to the mass of air that surrounds planets.
2. The troposphere is the first layer of this and is where weather occurs.
3. Earth's contains about 78.08% nitrogen, 20.95% oxygen, 0.93% argon, 0.038% carbon dioxide, trace amounts of other gases, and varying amounts of water vapor.

ATOMS
1. These are the smallest components of an element that can exist alone or in combination.
2. They are made up of protons, neutrons and electrons.
3. They are the smallest components of an element having the chemical properties of that element.

BIOLOGY
1. This science is concerned with the structure, function, distribution, adaptation, interactions, and evolution of plants and animals.
2. Botany is the branch that studies plant life. Zoology is the branch that studies animal life.
3. Genetics and Anatomy are two branches of this science.

CARNIVOROUS
1. It describes a predatory, flesh-eating animal.
2. It is an antonym of "herbivorous."
3. Lions, tigers, dogs and most humans are this.

CELL
1. It is the basic structural and functional unit in all living things.
2. It is the smallest structural unit of living matter able to function independently.
3. It is the basic microscopic unit of living things.

CHEMISTRY
1. This is the branch of the science dealing with the composition of substances and their properties and reactions.
2. You would find information about acids and bases in a textbook of this subject.
3. The Periodic Table is used in this branch of science.

COMPOUND(S)
1. A ___ is a distinct substance formed by chemical union of 2 or more ingredients. All ___ are molecules, but not all molecules are ___.
2. Water, carbon dioxide & methane are ___ because each is made from more than 1 element.
3. Carbon dioxide is a ___. Its formula is CO_2. It is composed of two oxygen atoms bonded to a single carbon atom.

General Science Bingo

CORE
1. Earth has an inner one and an outer one. The inner one is probably made of iron and nickel.
2. This is the name for the central part of a celestial body, such as a planet or moon.
3. This is the central portion of the Earth below the mantle.

CRUST
1. This layer of Earth is very thin compared to the other layers.
2. It is what we call Earth's surface.
3. Tectonic plates float on it.

DATA
1. It is a collection of facts from which conclusions may be drawn.
2. Sometimes graphs are used to organize this group of facts, statistics or measurements.
3. This factual information can be used as a basis for reasoning, discussion, or calculation.

EARTHQUAKES
1. They result from a sudden release of energy in the Earth's crust that creates seismic waves.
2. The point on the Earth's surface that is directly above the focus, or hypocenter, of one is called the epicenter.
3. In 1935 Charles F. Richter devised a scale to describe their intensity.

ECOLOGY
1. It is the study of how organisms interact with each other and their physical environment.
2. The study of biomes, which are communities characterized by the dominant forms of plant and animal life and the prevailing climate, are included in this branch of science.
3. The study of coral reefs and other ecosystems would be included in this branch of Earth Science.

ELECTRICITY
1. This form of energy is measured in units of power called watts and in units of force called volts.
2. Its rate of flow is measured in amperes.
3. Lightning is a static form of this.

ELEMENT
1. It is a substance that cannot be broken down by chemical means. They are defined by their number of protons.
2. The Periodic Table lists all of them. They are arranged according to atomic number.
3. Hydrogen is the most abundant one in the universe and makes up almost 75 % of all matter.

ENERGY
1. It is the capacity to do work. There are two main categories, potential and kinetic.
2. Fossil fuels, such as coal, oil and natural gas, are a non-renewable source of this.
3. Some forms of this include thermal, or heat; light; sound; chemical; and nuclear.

EROSION
1. It is the wearing away of Earth's surface by a natural process.
2. The main agent of this process is running water; other agents include glaciers, the wind, and waves breaking against the coast.
3. Unlike weathering, this process doesn't involve the chemical or physical breakdown of the rocks' minerals, just the carrying away of particles.

EXPERIMENT
1. It is the act of conducting a controlled test or investigation.
2. This is a method used in scientific inquiry to investigate particular types of problems.
3. This controlled test is sometimes made to examine the validity of a hypothesis.

General Science Bingo

FOOD CHAIN 1. This describes the feeding relationships between species within an ecosystem. 2. Plankton, microscopic plant life that floats in the ocean, is the first level of this. 3. It usually starts with a primary producer and ends with a carnivore.	**FREQUENCY** 1. It is a measure of the number of occurrences of a repeating event within a given time period. 2. In terms of a sound wave, this term refers to *how often* the particles in the air vibrate. 3. The shorter the wavelength, the higher the ___.
GENE 1. It is the basic biological unit of heredity. 2. It is a segment of DNA at a specific location on a chromosome. 3. This hereditary unit determines a particular characteristic in an organism.	**GEOLOGY** 1. This branch of science deals with the origin, history, and structure of the Earth. 2. Rocks and minerals would be studied in this branch of science. 3. Petrology, seismology and volcanology are subdivisions of this branch of science.
GRAVITY 1. It is the natural force of attraction between any two objects with mass. 2. It is what keeps our moon in Earth's orbit and the planets in orbit around the sun. 3. Newton's Law of ___ states that every particle in the universe attracts every other particle.	**HEAT** 1. This form of energy flows between two samples of matter because of their difference in temperature. The energy is created by the motion of the atoms and molecules. 2. This form of energy moves, spreading out from hotter things to cooler things. 3. This type of energy moves by conduction, convection and radiation.
HERBIVOROUS 1. These kind of animals feed on plants. 2. It is the opposite of "carnivorous." 3. Animals that are both carnivorous and this are said to be omnivorous.	**LIGHT** 1. It is the only form of energy that we can see directly. 2. This form of electromagnetic energy is visible to the human eye. 3. Green plants convert this type of energy into chemical energy through photosynthesis.
MAGNET 1. It is a material that produces a force which attracts iron, nickel or cobalt. 2. Lodestone is a natural one. 3. If one is held freely, it would line up with Earth's north and south poles.	**MAMMALS** 1. These warm-blooded vertebrates have hair. The females produce milk with which to nourish their young. 2. Monkeys, rabbits, bats, elephants, whales, and kangaroos are all examples. 3. Woolly mammoths and saber-toothed tigers are extinct examples.

MANTLE

1. It is a highly viscous layer directly under the crust and above the outer core.
2. The distinction between the crust and this layer is based on chemistry, rock types, flow and seismic characteristics.
3. This layer of Earth, which extends to a depth of about 1,800 miles, accounts for about 85% of the total weight and mass of the Earth.

MASS

1. This is the property that causes matter to have weight in a gravitational field. On Earth's surface it is the same as weight.
2. It is the amount of matter in something.
3. The SI unit of this is the kilogram (kg).

MATTER

1. It occurs in three main states, or phases: solid, liquid and gas.
2. Its fourth state, plasma, occurs when electrons are stripped away by high heat or pressure.
3. It changes from a solid state to a liquid state at its melting point.

MICROSCOPE

1. This optical instrument utilizes one or more lenses to magnify images of small objects.
2. An electron ___ uses electrons rather than visible light to produce magnified images.
3. This piece of laboratory equipment magnifies things too small to be seen by the naked eye or too small for their details to be seen by the naked eye.

MOLECULE

1. It is the simplest structural unit of an element or a compound and is made up of even smaller particles called atoms.
2. It is a unit of atoms bonded together. Some are compounds.
3. If it contains at least two different elements, then it is a compound.

MOTION

1. It is a continuous change in the location of a body as the result of applied force.
2. Newton's First Law of ___ is sometimes referred to as the Law of Inertia.
3. Newton's First Law states that an object in ___ tends to remain in this state unless an external force is applied to it.

NUCLEUS

1. In physics, it is the positively charged central region of an atom, composed of protons and neutrons.
2. This part of an atom contains most of its mass.
3. In biology, this part of the cell contains the chromosomes with the DNA and directs the cell's activities.

OCEANOGRAPHY

1. This branch of Earth Science is sometimes called marine science.
2. This branch of science covers ocean currents and waves; marine organisms and ecosystems; plate tectonics and the geology of the sea floor; and more.
3. Deep submergence vehicles are important to scientists who specialize in this field of science.

PHOTOSYNTHESIS

1. This is the process by which green plants use water and carbon dioxide to create glucose and oxygen.
2. This process takes energy from the sun and converts it into a storable form which plants use for their own life processes.
3. Animals provide the carbon dioxide needed for this process and get oxygen in return.

PHYSICS

1. This is the science of matter and energy and their interactions.
2. Fields within this science include acoustics, optics, mechanics, and electromagnetism.
3. Motion, force and simple machines are concepts in this science.

General Science Bingo

PLANET(S) 1. Earth is 1 of 4 terrestrial ____. All have the same basic structure: a central metallic core, mostly iron; a silicate mantle; and a crust with canyons, craters, mountains, and volcanoes. 2. This kind of celestial body orbits a star and is massive enough to be rounded by its own gravity. 3. Terrestrial ones are composed mostly of silicate rocks; they are also called rocky ____.	**RADIATION** 1. This term is used to describe types of energy that come from a source and travel through some material or through space. 2. Examples are light, heat and sound. 3. Heat waves, sound waves and light rays are emissions of this type of energy.
SCIENTIFIC METHOD 1. This process is the basis for scientific inquiry. 2. This process follows a series of steps: identify a problem, formulate a hypothesis, test the hypothesis, collect and analyze the data, and make conclusions. 3. This procedure for analyzing scientific problems in a way that leads to verifiable results is based upon controlled experiments.	**SOLAR SYSTEM** 1. Ours comprises our sun; the planets and their satellites; and the other celestial bodies that orbit the sun. 2. In general, it refers to any star and the bodies that orbit it. 3. The sun is at the center of it.
SOUND 1. The number of vibrations per second is the frequency of the ____. 2. This type of energy travels more slowly than light energy. 3. The more quickly something vibrates, the higher the ____.	**TEMPERATURE** 1. As used in chemistry and other physical sciences, this is a measurement of the average kinetic energy in a sample—in other words, how fast the molecules are vibrating. 2. Three scales used to measure this value are Kelvin, Celsius, and Fahrenheit. 3. We commonly refer to it as the degree of hotness or coldness of a body or environment.
THEORY 1. A scientifically acceptable general principle or body of principles explaining a phenomenon is often called a ____. 2. It is an explanation or model based on observation, experimentation, and reasoning, especially one that has been tested. 3. Albert Einstein is known for his ____ of Relativity.	**TIDES** 1. This is the rising and falling of Earth's oceans and other waters caused by gravitational forces of the moon and to a lesser extent the sun. 2. Neap ____ are weak and occur between the first and third quarters of the moon 3. Spring ____ occur near the times of the new moon and full moon and have the most variation in water level.
VOLCANOES 1. ____ are vents in Earth's surface through which magma and associated gases and ash erupt; they are also the structures that are created as a result. 2. The eruption of low-viscosity lavas that flow far from the vent result in shield ____. 3. Some are built by the piling up of ejected fragments around the vent in the shape of a cone with a central crater. General Science Bingo	**WATER CYCLE** 1. This describes the continuous movement of water on, above, and below the surface of the Earth. 2. It is also called the hydrologic cycle. 3. The sun is an important part of this because its heat evaporates water from the oceans into the atmosphere to form clouds.

General Science Bingo

Element	Acids	Biology	Geology	Magnet
Astronomy	Anatomy	Temperature	Mantle	Tides
Atmosphere	Water Cycle		Microscope	Core
Radiation	Electricity	Volcanoes	Heat	Matter
Motion	Experiment	Physics	Theory	Solar System

General Science Bingo: Card No. 1

General Science Bingo

Radiation	Astronomy	Mammals	Scientific Method	Gene
Matter	Ecology	Atmosphere	Cell	Herbivorous
Crust	Experiment		Energy	Volcanoes
Oceanography	Molecule	Water Cycle	Photosynthesis	Solar System
Tides	Temperature	Physics	Carnivorous	Theory

General Science Bingo: Card No. 2

General Science Bingo

Radiation	Volcanoes	Ecology	Heat	Astronomy
Mantle	Anatomy	Compound(s)	Acids	Cell
Electricity	Temperature		Herbivorous	Atoms
Water Cycle	Crust	Motion	Oceanography	Mammals
Theory	Carnivorous	Physics	Photosynthesis	Gene

General Science Bingo

Water Cycle	Herbivorous	Biology	Cell	Gene
Light	Chemistry	Acids	Scientific Method	Astronomy
Microscope	Oceanography		Magnet	Geology
Volcanoes	Data	Temperature	Physics	Atmosphere
Carnivorous	Tides	Molecule	Theory	Core

General Science Bingo

Tides	Magnet	Electricity	Atmosphere	Carnivorous
Light	Volcanoes	Compound(s)	Energy	Anatomy
Biology	Core		Mantle	Frequency
Solar System	Gene	Element	Photosynthesis	Earthquakes
Ecology	Physics	Astronomy	Water Cycle	Microscope

General Science Bingo

Atoms	Herbivorous	Mammals	Gene	Core
Heat	Electricity	Earthquakes	Acids	Astronomy
Scientific Method	Cell		Chemistry	Energy
Physics	Motion	Photosynthesis	Molecule	Biology
Matter	Atmosphere	Element	Microscope	Data

General Science Bingo

Element	Herbivorous	Frequency	Mantle	Ecology
Matter	Gene	Experiment	Anatomy	Light
Mammals	Geology		Energy	Chemistry
Water Cycle	Oceanography	Compound(s)	Radiation	Crust
Physics	Carnivorous	Photosynthesis	Molecule	Atoms

General Science Bingo

Microscope	Herbivorous	Erosion	Heat	Chemistry
Light	Biology	Scientific Method	Core	Atmosphere
Data	Nucleus		Gene	Magnet
Theory	Water Cycle	Radiation	Cell	Oceanography
Temperature	Physics	Molecule	Electricity	Matter

General Science Bingo: Card No. 8

© Barbara M. Peller

General Science Bingo

Energy	Ecology	Experiment	Data	Carnivorous
Cell	Photosynthesis	Microscope	Electricity	Herbivorous
Gravity	Element		Anatomy	Erosion
Earthquakes	Solar System	Motion	Mantle	Frequency
Oceanography	Gene	Compound(s)	Radiation	Magnet

General Science Bingo

Radiation	Heat	Chemistry	Scientific Method	Data
Core	Atmosphere	Acids	Anatomy	Gene
Nucleus	Herbivorous		Geology	Crust
Motion	Solar System	Earthquakes	Photosynthesis	Gravity
Compound(s)	Matter	Mammals	Tides	Microscope

General Science Bingo

Atoms	Herbivorous	Electricity	Earthquakes	Matter
Erosion	Gravity	Mantle	Energy	Acids
Light	Gene		Mammals	Experiment
Compound(s)	States' Rights	Photosynthesis	Carnivorous	Radiation
Cell	Physics	Element	Molecule	Ecology

General Science Bingo

Ecology	Magnet	Gravity	Heat	Energy
Experiment	Matter	Biology	Molecule	Anatomy
Element	Frequency		Core	Scientific Method
Physics	Oceanography	Gene	Radiation	Light
Herbivorous	Erosion	Nucleus	Cell	Atmosphere

General Science Bingo

Earthquakes	Magnet	Atoms	Gravity	Core
Biology	Erosion	Gene	Energy	Crust
Heat	Ecology		Experiment	Frequency
Microscope	Photosynthesis	Chemistry	Nucleus	Radiation
Physics	Solar System	Molecule	Element	Mantle

General Science Bingo

Carnivorous	Gene	Electricity	Energy	Cell
Atmosphere	Element	Gravity	Anatomy	Herbivorous
Earthquakes	Geology		Mammals	Compound(s)
Solar System	Photosynthesis	Nucleus	Chemistry	Atoms
Physics	Scientific Method	Crust	Matter	Microscope

General Science Bingo

Mantle	Energy	Electricity	Ecology	Heat
Atoms	Mammals	Acids	Biology	Cell
Core	Element		Astronomy	Herbivorous
Physics	Gravity	Erosion	Photosynthesis	Earthquakes
Matter	Oceanography	Molecule	Data	Experiment

General Science Bingo

Chemistry	Gravity	Erosion	Data	Planet(s)
Scientific Method	Crust	Frequency	Light	Geology
Earthquakes	Magnet		Core	Experiment
Water Cycle	Atmosphere	Physics	Mass	Radiation
Cell	Sound	Molecule	Oceanography	Herbivorous

General Science Bingo: Card No. 16

© Barbara M. Peller

General Science Bingo

Compound(s)	Mass	Food Chain	Gravity	Carnivorous
Mantle	Cell	Photosynthesis	Geology	Frequency
Energy	Microscope		Sound	Erosion
Solar System	Matter	Radiation	Electricity	Crust
Motion	Earthquakes	Ecology	Heat	Magnet

General Science Bingo

Data	Nucleus	Atmosphere	Earthquakes	Scientific Method
Herbivorous	Compound(s)	Motion	Core	Cell
Energy	Crust		Food Chain	Biology
Solar System	Acids	Photosynthesis	Radiation	Mammals
Sound	Gravity	Electricity	Mass	Atoms

General Science Bingo

Core	Atoms	Gravity	Erosion	Radiation
Mantle	Heat	Herbivorous	Ecology	Geology
Mass	Carnivorous		Anatomy	Astronomy
Mammals	Sound	Motion	Oceanography	Food Chain
Biology	Planet(s)	Matter	Microscope	Molecule

© Barbara M. Peller

General Science Bingo

Nucleus	Mass	Heat	Gravity	Molecule
Atmosphere	Experiment	Light	Motion	Scientific Method
Magnet	Frequency		Water Cycle	Acids
Tides	Temperature	Theory	Oceanography	Sound
Volcanoes	Microscope	Planet(s)	Radiation	Food Chain

© Barbara M. Peller

General Science Bingo

Mantle	Atoms	Light	Gravity	Tides
Magnet	Food Chain	Chemistry	Erosion	Element
Crust	Matter		Mass	Electricity
Motion	Ecology	Sound	Solar System	Microscope
Water Cycle	Planet(s)	Molecule	Compound(s)	Oceanography

General Science Bingo

Data	Mammals	Food Chain	Biology	Earthquakes
Scientific Method	Heat	Astronomy	Erosion	Anatomy
Atmosphere	Geology		Element	Frequency
Sound	Solar System	Oceanography	Acids	Carnivorous
Planet(s)	Compound(s)	Mass	Crust	Light

General Science Bingo: Card No. 22

General Science Bingo

Chemistry	Mass	Ecology	Biology	Molecule
Atoms	Nucleus	Matter	Mantle	Acids
Mammals	Earthquakes		Theory	Element
Crust	Planet(s)	Sound	Compound(s)	Oceanography
Tides	Temperature	Microscope	Motion	Food Chain

© **Barbara M. Peller**

General Science Bingo

Chemistry	Nucleus	Carnivorous	Mass	Erosion
Food Chain	Molecule	Light	Scientific Method	Element
Frequency	Data		Earthquakes	Crust
Tides	Theory	Sound	Compound(s)	Magnet
Volcanoes	Water Cycle	Planet(s)	Heat	Temperature

© Barbara M. Peller

General Science Bingo

Water Cycle	Light	Mass	Electricity	Food Chain
Acids	Solar System	Mantle	Chemistry	Anatomy
Magnet	Erosion		Theory	Sound
Astronomy	Tides	Temperature	Planet(s)	Geology
Molecule	Carnivorous	Atmosphere	Cell	Volcanoes

General Science Bingo

Food Chain	Mass	Mammals	Scientific Method	Data
Motion	Heat	Erosion	Nucleus	Chemistry
Solar System	Theory		Geology	Water Cycle
Compound(s)	Biology	Tides	Planet(s)	Sound
Frequency	Cell	Electricity	Temperature	Volcanoes

General Science Bingo: Card No. 26

© Barbara M. Peller

General Science Bingo

Mammals	Atmosphere	Mass	Nucleus	Experiment
Tides	Theory	Mantle	Sound	Anatomy
Photosynthesis	Temperature		Planet(s)	Water Cycle
Data	Atoms	Light	Volcanoes	Acids
Cell	Geology	Food Chain	Astronomy	Frequency

General Science Bingo: Card No. 27

General Science Bingo

Core	Nucleus	Astronomy	Mass	Chemistry
Experiment	Food Chain	Theory	Scientific Method	Geology
Temperature	Crust		Frequency	Motion
Radiation	Data	Matter	Planet(s)	Sound
Biology	Energy	Cell	Volcanoes	Tides

General Science Bingo: Card No. 28

General Science Bingo

Food Chain	Nucleus	Data	Mantle	Energy
Solar System	Motion	Light	Frequency	Astronomy
Magnet	Theory		Anatomy	Mass
Electricity	Tides	Gene	Planet(s)	Sound
Cell	Erosion	Volcanoes	Atoms	Temperature

© Barbara M. Peller

General Science Bingo

Carnivorous	Mass	Scientific Method	Energy	Sound
Acids	Nucleus	Mammals	Geology	Anatomy
Solar System	Earthquakes		Frequency	Light
Volcanoes	Atoms	Biology	Planet(s)	Theory
Tides	Ecology	Temperature	Food Chain	Astronomy

General Science Bingo: Card No. 30

www.ingramcontent.com/pod-product-compliance
Lightning Source LLC
Chambersburg PA
CBHW051419200326
41520CB00023B/7292